DISASTROUS DEATHS

Death by Atrocious Animals

by Mignonne Gunasekara

BEARPORT PUBLISHING

Minneapolis, Minnesota

Library of Congress Cataloging-in-Publication Data

Names: Gunasekara, Mignonne, author.
Title: Death by atrocious animals / Mignonne Gunasekara.
Description: Minneapolis,MN : Bearport Publishing, 2021. | Series: Disastrous deaths | Includes bibliographical references and index.
Identifiers: LCCN 2020058663 (print) | LCCN 2020058664 (ebook) | ISBN 9781636911687 (library binding) | ISBN 9781636911731 (ebook)
Subjects: LCSH: Animal attacks–Juvenile literature. | Biography–Miscellanea–Juvenile literature. | Death–Miscellanea–Juvenile literature.
Classification: LCC QL100.5 .G86 2021 (print) | LCC QL100.5 (ebook) | DDC 591.5/3–dc23
LC record available at https://lccn.loc.gov/2020058663
LC ebook record available at https://lccn.loc.gov/2020058664

© 2022 Booklife Publishing
This edition is published by arrangement with Booklife Publishing.

North American adaptations © 2022 Bearport Publishing Company. All rights reserved. No part of this publication may be reproduced in whole or in part, stored in any retrieval system, or transmitted in any form or by any means, electronic, mechanical, photocopying, recording, or otherwise, without written permission from the publisher.

For more information, write to Bearport Publishing, 5357 Penn Avenue South, Minneapolis, MN 55419. Printed in the United States of America.

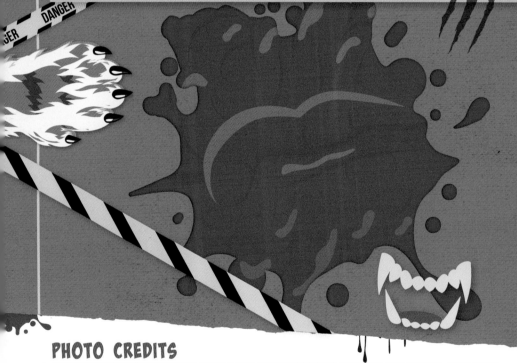

PHOTO CREDITS

All images are courtesy of Shutterstock.com, unless otherwise specified. With thanks to Getty Images, Thinkstock Photo, and iStockphoto. Background texture throughout - Abstracto. Gravestone throughout - MaryValery. Front Cover - ONYXprj. 5 - dptro, YummyBuum, Nipun Kundu. 6 - hvostik, Cosmic_Design, Panptys, Slowga. 7 - Anastasios71, ONYXprj. 8 - vangelis aragiannis, Viacheslav Lopatin, Darth_Vector. 9 - ariy, Lefteris Papaulakis, Eroshka, matrioshka. 10 - Strilets, bartamarabara, Maglara. 11 - Vertyr. 12 - Nicku, National Library of Scotland [CC0], Oxy_gen, Paragorn Dangsombroon. 13 - Royal Air Force official photographer [Public domain], Bodor Tivadar, Maquiladora, Anatolir, ArtMalivanov. 14 - Dim Tik. 15 - Pavel L Photo and Video, 1507kot. 16 - Greenshed [Public domain], Artist: D'Aveline (French artist, late 17th and early 18th century) [Public domain]. 17 - Public Domain, https://commons.wikimedia.org/w/index.php?curid=584067, Anan Kaewkhammul, Maike Hildebrandt, HappyPictures. 18 - andrey oleynik. 19 - PsyComa. 20 - uzuri, Everett Historical, Maquiladora. 21 - Charles Chusseau-Flaviens [Public domain], The Official Site of the Greek Royal Family [Public domain], Milan M. 23 - delcarmat, Maquiladora, gubernat. 24 - vchal, Plotitsyna NiNa. 25 - Hoika Mikhail, Everett - Art. 26 - LOVE YOU. 27 - Peter Maerky. 28 - V_E, Svetlana Foote, MarGi, svtdesign. 29 - Sarah2, Pogorelova Olga, Dwra.* - U.S. work public domain in the U.S. for unspecified reason but presumably because it was published in the U.S. before 1924.
Additional illustrations by Jasmine Pointer.

CONTENTS

Welcome to the Disaster Zone 4
Aeschylus 6
To Die, or Not to Die 8
Eleazar Avaran 10
Beasts in Battle 12
Hannah Twynnoy 14
Crouching Tiger,
   Annoying Hannah 16
Alexander I of Greece 18
It's Just a Flesh Wound 20
Heraclitus 22
Dropsy, Dung, and Dogs 24
Philip of France 26
Sanitation Situation 28

Timeline of Deaths 30
Glossary 31
Index 32
Read More 32

WELCOME TO THE DISASTER ZONE

History is full of grisly stories, weird tales, and a lot of death. Some of the more bizarre deaths of people involve animals. From covering themselves in cow poop and being **devoured** by wild dogs to being crushed by falling elephants and tortoises, people have found some pretty creative ways to get themselves killed.

Since the beginning of human history, about 107 billion people have lived on Earth. You know what that means . . . there are plenty of deaths to choose from!

In this book, we are going to look at the stories of six people who were taken out by atrocious animals, whether it was a **stampeding** pig, monkeys gone mad, or a taunted tiger having a tantrum.

Into the Disaster Zone We Go . . .

Throughout history, there have been lots of strange sayings that mean someone has died.

Here are a few of the weird ones:

- Kicked the bucket
- Bit the dust
- Met their maker
- Six feet under
- Food for worms
- Pushing up daisies

5

AESCHYLUS

Aeschylus was a famous **playwright** in ancient Greece. He had a terrible fear of being killed by falling objects. For this reason, he stayed outside as much as possible, thinking nothing could possibly fall out of the sky and hit him.

But one day, an eagle mistook Aeschylus's shiny bald head for a rock and dropped a tortoise on him. Eagles do this to crack open tortoises' shells and eat their insides. We don't know if the tortoise survived the incident, but we do know that poor Aeschylus didn't.

Aeschylus is sometimes known as the father of **tragedy** because he wrote plays with endings full of sadness and death—not unlike his own ending!

BORN: Eleusis, Greece
DIED: Gela, Italy
CLAIM TO FAME: Father of tragedy
DEATH BY: Tumbling tortoise

Oh the tragedy of going bald!

525–456 BCE

TO DIE, OR NOT TO DIE

Aeschylus is said to have written around 90 plays, but only 7 of his tragedies are still around today.

Aeschylus

Aeschylus's Influence

Aeschylus changed how plays were written and performed. Before him, Greek plays had only one main actor and a group of actors who reacted to the main actor through song or dance.

Aeschylus's plays may have been performed at theaters like this one in Athens.

The main actor used to play every character in the story, wearing a different mask for each one. Aeschylus was the first to bring in a second main actor. Now the two main actors could talk to each other and have a conversation. This added more drama to the story.

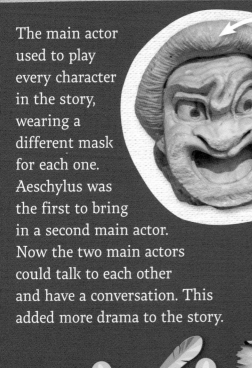

Ancient Greek theater masks

Aeschylus is also remembered for making costumes and stage scenery much more detailed.

Aeschylus wanted the audience to think about the characters. Were their actions right or wrong? Did they get treated fairly?

Aeschylus wrote plays about characters from Greek **mythology**.

ELEAZAR AVARAN

Eleazar Avaran had joined the **rebellion** against the Greek King Antiochus V in the second century BCE. Antiochus was trying to destroy the Jewish religion and make everyone **worship** Greek gods instead. He had seized Jerusalem, the religious and political center of Jewish life. Jewish rebels fought to regain control of the city and drive out the forces of Antiochus.

During the battle, Eleazar thought he spotted the war elephant that the Greek king was riding. Eleazar rushed over and stuck his spear into the elephant's belly ... but the elephant collapsed on top of him and crushed him to death.

Eleazar's side didn't win this battle, but the rebellion is important to Jewish people and is remembered during **Hanukkah**.

BORN: Unknown
DIED: Beth Zecheriah, Jerusalem
CLAIM TO FAME: Gutsy in battle
DEATH BY: Elephant error

I didn't think this all the way through . . .

DIED
163 BCE

BEASTS IN BATTLE

For centuries, animals of all sizes have been used in war. Horses were often used to carry food, medicine, weapons, and injured soldiers.

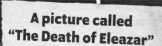

A picture called "The Death of Eleazar"

Dogs were trained to find wounded soldiers and carry medicine to them on the battlefield. Dogs have also been used to sniff out hidden bombs.

A World War I military dog

Before World War I (1914–1918), horses were used to carry soldiers and pull **artillery** and equipment. After the first world war, horses were replaced by tanks.

Private... Pigeon?

Carrier pigeons were used to send messages, especially to and from the battlefield. More than 9 out of 10 messages sent by pigeon during World War I were delivered successfully.

A member of the Royal Air Force holding a carrier pigeon

A pigeon called Cher Ami was awarded a medal for her services during World War I. She delivered her last message even after being shot. The message saved the lives of nearly 200 American soldiers.

Carrying such important information was dangerous. Some pigeons even became prisoners of war!

A carrier pigeon

HANNAH TWYNNOY

A woman named Hannah Twynnoy worked at the White Lion pub in Malmesbury, England. One day in 1703, a traveling **menagerie** came to town. One of its attractions was a tiger. The story goes that Hannah kept annoying the tiger—even after she was warned not to. At some point, the tiger managed to escape and ended up attacking Hannah. How does the story end? Let's just say Hannah never bothered that tiger again.

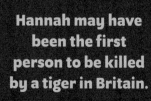

Hannah may have been the first person to be killed by a tiger in Britain.

CROUCHING TIGER, ANNOYING HANNAH

A poem on Hannah's gravestone reads:

Hannah's gravestone

In bloom of Life
She's snatchd from hence,
She had not room
To make defence;
For Tyger fierce
Took Life away.
And here she lies
In a bed of Clay,
Until the Resurrection Day.

MENAGERIE MADNESS

People have used animals for entertainment throughout history. The ancient Romans used to make animals fight and perform at the **Colosseum.** Forcing bears to fight dogs was popular in sixteenth-century London.

French king Louis XIV's seventeenth-century menagerie

Wealthy people often had menageries. They collected **exotic** and rare animals as **status symbols.**

Traveling menageries became popular in the eighteenth and nineteenth centuries as more people traveled around the world. These voyagers brought back new and interesting animals. The menageries developed into the circuses and zoos we have today.

A painting of London Zoo, 1835

The Tower of London's Royal Menagerie was started in 1235. It included lions, an elephant, a polar bear, and a leopard.

At first, menageries just showed animals in cages, but people eventually became bored. The animals were then forced to perform to keep audiences interested.

ALEXANDER I OF GREECE

Alexander I came to rule Greece in 1917 when he was 24 years old. He had been king for only three years when something very strange—and tragic—happened. Alexander was walking his dog, Fritz, when Fritz was attacked by a pair of monkeys. Alexander tried to break up the fight, and the monkeys turned on him.

Alexander was bitten. At first, he didn't think it was very serious. Over the next few days, however, an **infection** spread throughout his body and killed him.

The monkeys that attacked Fritz and Alexander were a pair of Barbary macaques that belonged to a palace worker.

BORN: Athens, Greece
DIED: Tatoi, Greece
CLAIM TO FAME: King of Greece, 1917–1920
DEATH BY: Mad monkeys

Don't make a fuss—it's just a scratch!

1893-1920

19

IT'S JUST A FLESH WOUND

Little did anyone realize that the attack from these Barbary macaques would change history.

A Barbary macaque

Following Alexander's death, his father, Constantine I, was brought back to rule Greece. Previously, Constantine had been forced to leave the throne during World War I because he refused to join the Allies. Alexander had been happy to fight with the Allies.

Allied soldiers during World War I

The Allied powers of World War I fought together against Germany and its supporters. Allied countries included Britain, France, Russia, Italy, and Japan.

To reward Alexander for fighting on their side, the Allies offered to help Greece expand its borders into Turkey. Things had been going well for Greece . . . until Alexander died and Constantine returned.

Alexander I

Constantine I

With Constantine back in power, the Allies no longer wanted to help Greece. Without assistance, Greece could not fight the Turkish forces defending their territory. Greece was defeated and driven away.

British Prime Minister Winston Churchill blamed the two monkeys for all the Greek and Turkish bloodshed.

A quarter of a million people died because of this monkey's bite.

HERACLITUS

Heraclitus was a Greek **philosopher** who lived alone in the mountains. In the year 480 BCE, Heraclitus became sick and believed the cure would be to cover himself in cow dung!

He hoped that the warmth of the fresh dung would draw the illness out of his body. However, the dung soon dried out and became a solid cast around his body—Heraclitus was trapped in poop! Then, a pack of hungry wild dogs appeared and eagerly snacked on the poop-wrapped philosopher.

Heraclitus is known as the Weeping Philosopher because he was so sad. He also earned the name the Dark Philosopher because his writings were so hard to understand.

BORN: Ephesus, Turkey
DIED: Ephesus, Turkey
CLAIM TO FAME: Sad philosopher
DEATH BY: Hungry wild dogs

"Laugh it up! I tell you, this will work!"

540–480 BCE

23

DROPSY, DUNG, AND DOGS

Heraclitus had a condition called dropsy, which is when water collects in parts of the body where it doesn't belong. It can be very painful, which may explain why Heraclitus was not thinking clearly.

Dropsy is now known as edema.

MR. KNOW-IT-ALL

The ruins of the Temple of Artemis at Ephesus

Heraclitus wrote all his ideas down in one book, which he left at the Temple of Artemis at Ephesus. The book was mostly lost—just over 100 **fragments** of it are still around today.

24

One of Heraclitus's most famous ideas was that things are always changing. He called this **flux**.

Artemis

Artemis is the Greek goddess of the moon, the hunt, and wild animals. Maybe she should have told those wild dogs to sit and stay!

A painting of Heraclitus

Another idea of Heraclitus was that opposites—such as hot and cold—were closely related to each other and created balance in the universe.

25

PHILIP OF FRANCE

In 1129, Philip was made joint king of France alongside his father Louis VI. Philip was only 13 years old at the time, but Louis had high hopes that Philip would be a good king.

However, Philip didn't get a chance to prove his father right. Just two years after rising to the throne, Philip and a group of friends were riding their horses through the streets of Paris. Suddenly, a black pig appeared out of a dung heap in front of them. Philip's horse tripped over it, and the young king was sent flying from his saddle. Philip died of his injuries.

Philip wasn't the best-behaved person during his short reign. He was very rude and he refused to listen to advice.

SANITATION SITUATION

Why were there piles of poo lying in the road in Paris in 1131? Without indoor toilets, poop was pretty much everywhere back then!

At that time, a popular way to get rid of waste was to dump it in rivers. This was a bad idea because people used the water from those same rivers to cook and clean. This often spread illness and disease.

Public toilets on London Bridge emptied waste into the river Thames below. *Yuck!*

Waste would also be dumped into the street in the hopes that rain would wash it away. Anything that was left would be cleaned up by workers known as muck-rakers.

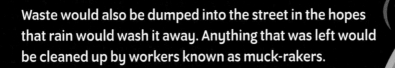

Muck-rakers picked up waste from the streets and took it to dumps outside the city.

The Great Stink

In the summer of 1858, the hot sunshine baked all the poop that had been dumped into the Thames for years, and the smell was unbearable. Politicians were so grossed out that they ordered a new sewer system to be built in London.

Chamber pots were kept in bedrooms. People would poop and pee in them and empty them into gutters or **cesspits**.

TIMELINE OF DEATHS

Heraclitus
480 BCE

Aeschylus
456 BCE

Eleazar Avaran
163 BCE

King Philip
1131

Hannah Twynnoy
1703

Alexander I
1920

GLOSSARY

artillery large guns used to shoot over a great distance

cesspits holes in the ground that collect human waste

Colosseum a huge theater built in ancient Rome

devoured quickly and completely eaten

exotic from a new or foreign place

flux continuous change or a series of changes

fragments small pieces that have been broken off or separated from a whole

Hanukkah an eight night Jewish holiday that celebrates the Temple of Jerusalem being taken back from Antiochus IV

infection when a wound or body part is diseased because bacteria or a virus has gotten inside it

menagerie a collection of animals, usually for entertainment or research purposes

mythology stories that tell of gods and heroes

philosopher a person who studies the nature of knowledge, reality, and life

playwright a person who writes plays

rebellion when people fight back against the rules of somebody who is in control

stage scenery items and backgrounds used on stage to make it look like a certain time or place

stampeding running in a wild and uncontrolled way

status symbols things that show how rich or important someone is

tragedy plays that are about something sad with no happy ending

worship honor or show respect for something

INDEX

audiences 9, 17
dogs 4, 12, 16, 18, 22–23, 25
eagles 6
elephants 4, 10–11, 17
fathers 6–7, 20, 26
horses 12, 26
illness 22, 28
kings 10, 16, 18–20, 26–27
menageries 14, 16–17
monkeys 5, 18–19, 21
muck-rakers 29
philosopher 22–23
pigeons 13
pigs 5, 26–27
poop 4, 22, 27–29
rivers 28
tigers 5, 14–15
toilets 28
tortoises 4, 6–7
war 10, 12–13, 20

READ MORE

Joyce, Markovics. *Deadly Venomous Mammals! (Envenomators).* Minneapolis: Bearport Publishing, 2019.
Mattern, Joanne. *Deadly Weapons (Earth's Amazing Animals: Animal Superpowers).* South Egremont, MA: Red Chair Press, 2019.
Ventura, Marne. *Animal Attacks (Surviving).* New York: AV2, 2019.